In Pastures Green

In Pastures Green

Parables From My Sheep Farm

Bev Condy

JAN WHITE

FR

Review and Herald® Publishing Association
Washington, DC 20039-0555
Hagerstown, MD 21740

This book was
Edited by Gerald Wheeler
Designed by Bill Kirstein
Cover design by Stephen Hall
Cover art by Isadore Seltzer
Inside illustrations by Jan White
Type set: 11.8 pt. Windsor Old Style Light

PRINTED IN U.S.A.

94 93 92 91 90   10 9 8 7 6 5 4 3 2 1

Library of Congress Cataloging-in-Publication Data

Condy, Bev. 1935-
    In pastures green: parables from my sheep farm / Bev. Condy.
p.   cm.
    1. Christian life—Seventh-day Adventist authors. 2. Condy, Bev.
1945- . 3. Sheep-Religious aspects—Christianity. 4. Country life—
Religious aspects—Christianity. 5. Sheep ranches—California.   I. Title.
BV4501.2.C6485 1990
242—dc20   89-24213
                                                                CIP

ISBN 0-8280-0543-5

Contents

> "For you were
> like sheep going astray,
> but now you have returned
> to the Shepherd and Overseer
> of your souls" *(1 Peter 2:25)*.

Introduction

Farm stories are not common today, nor are we attuned to God's voice in nature as in past generations. What special meaning can they have in our fast-paced, noisy society, rushing into the twenty-first century where listening is a lost art and communication break-down is rampant in homes, churches, and society at large.

In Scripture, however, because of the agrarian society of the times, Jesus used pastoral and agriculturally-based parables to speak to the people. Even then His disciples many times did not understand His parables and He would have to take them aside and explain each story to them in private.

Living on a small farm in California I have not always understood the stories I have seen played out in the daily activities on it. I had to go to Jesus privately and hear His voice and explanation for me in prayer.

Not all of us live on a farm. But we all have a sphere of influence, a place where we live and dwell. The people we know, the lives that we touch, that is our personal pasture. And that is exactly where God will feed you as He speaks if you will look and listen.

These simple, everyday stories, gleaned and compiled from a number of years of farm life, happened within the circle of my life, the sphere of my influence, and illustrate my own personal pasture. The prayers are what I heard God say to me as I lived through each drama.

My earnest prayer for you as you read this book is that He will help you to become aware that we all have pastures and stories to tell of His great love.

Oak Haven Farm

Truly, our farm was heaven-sent. The longer we live in our two-story country home atop the knoll, nestled among live oak trees and grassy fields, the more aware I become of how God leads us in our lives, right down to the perfect timing of our move from city to country.

It all began like this . . .

I had spent most of my life trying to be a "city girl." Until I met Jesus I had climbed up and down all kinds of social ladders, rung by rung. He freed me from that unpleasant exercise and showed me, ever so slowly, that I was a true country girl at heart.

One warm spring Sunday afternoon in 1970 my husband and I took a drive. We stopped at a real estate office, and on impulse Chuck and I signed the papers for ten beautiful, oak-studded acres in the lower foothills of the Sierra Nevada mountain range in northeastern California. It took us four years to begin to build, but we finally moved into our country house in October, 1975. However, it required even more time before it really became home for me.

I'll never forget that fall morning I awoke following our first night's sleep in our new home. Gaze fixed on the ceiling, I thought to myself, "Where on earth am I?"

We had moved, and I didn't even know where the local supermarket was. Disliking change, I longed for my favorite store down on the corner where I could find familiar items without a second glance. Now we were miles from any shopping center.

The move had been traumatic, a source of pain for me as I had to leave behind my friends, a beautifully landscaped yard with swimming pool, and, of course, my "ladder." All this to move to the country. At the time I thought it too much to ask of anyone, and felt resentful toward my husband for asking it of me. I felt like a combination of Lot's wife and the children of Israel longing for their leeks and onions. The Lord, however, began to speak quietly to me in the beauty of nature surrounding us. The miracle of it was that I could hear Him.

All our friends traveled the long distance to visit, so we had house guests by the score. Family and friends loved our country place, but in some ways the move had made our "empty nest" even more apparent. Our daughter and her husband lived far off in Alaska, our son was away at college, and I needed something to mother.

For my birthday the following spring my husband purchased two adorable orphan goats. They were just what we needed to eat the abundant spring grasses. By the time Chuck completed building the "goat hotel," we realized that the two wethered goats (neutered males) could never eat all that extra pasture. For a time I thought Arabian mares provided the answer. With the right breeding they could add to our income and also keep the grass cropped. Horses and riding had been a life-long hobby. But the cost of a mare was prohibitive, and I knew very little about other farm animals. Naturally I didn't want something that would end up being slaughtered. Then a friend suggested one day that if I raised black wool sheep she would buy the wool. A spark ignited, and my adventure of raising black Lincoln-Corriedale cross-bred sheep had begun.

I knew the Lord had a large part to play in the chain of events leading us to the country, because now I was hearing Him speak to me more clearly in my morning devotions, during jogs, and throughout my everyday activities. His book of nature had become a lesson book for me. God—knowing my heart and my love for people and animals more than I did—chose these methods to enable me to hear His voice more intimately.

A book by Phillip Keller, *A Shepherd Looks at Psalm 23,* spoke to my heart. In the chapter titled, "He Maketh Me to Lie Down in Green Pastures," I saw myself as the ewe who was discontent with her pasture. Learning from my sheep that they will not rest until they are content, I longed to be content myself, for I recognized that God's providence had brought us here. He had gently placed me in this country pasture where He could feed me and I could hear His voice.

JaN·WHITE

OUTDOOR PASTURES

"Trust in the Lord and do good, dwell in the land and enjoy safe pasture. Delight yourself in the Lord and He will give you the desires of your heart" (Psalm 37:3, 4).

Pig Pen

THE spring grasses were rapidly turning to summer brown. My recently acquired ewes had grown discontented in the confines of our ten acres. They had found a broken spot in the rail fence and had made their escape, meandering down to my neighbor's back pasture, through her fence, and onto her property. The silence of their bells rang an alarm in my mind.

Responding to my new role as a shepherd, I stopped what I was doing and made a hasty survey of my acreage. To my dismay, I did not find the sheep anywhere. After climbing over the back fence I found the hole and followed the bent grass path the lost sheep had made. The sound of their tinkling bells eased my mind and directed my feet.

Then I saw them, huddled together with heads hanging sheepishly down, in a dry, hot, barren, vacant pig pen! They had no water and looked miserable, trapped in a pen with an open gate.

"They got in there all by themselves. Certainly they can come out on their own," I thought as I climbed over the neighbor's wooden fence to rescue the ewes. But they didn't.

"Sheep are rather dumb," I concluded. However, when they saw me they greeted me with happy sounds, and I led them out through the open gateway. Then they took off running back to the hole in our fence line, where they hopped through it and scampered with eager anticipation down to

13

the big oak watering trough. Their thirst was matched by their delight to be home.

"Why is it, Lord, that when I break through the safety of Your will for me, I rush head on into dry and barren lands not even fit for pigs? There is no water of your Holy Spirit for my soul, and I stand numbed by my choice. I am so thankful You come after me and help me see that the door of mercy stands open. You gently bring me back to the safety of Your fold, the fence line of Your commandments. Thank you that the Water of Life flows freely, and as I eagerly come to drink, may praise and adoration be heard from my lips as I share the joy that the Lord is my Shepherd." ♦

New Pasture

IT was early fall. We had cleared the new pasture in preparation for the arrival of our ram and placed our ewes in it so they could adjust. Being a novice shepherd, I was not aware that sheep are creatures of habit and dislike change. I had no idea they would spend all day staring longingly at the pasture they had just left, their eyes filled with fear and insecurity. They stood together in a woolly clump with their heads to the ground.

The ram arrived after nightfall and the next morning I awoke with eager anticipation, hoping to see my ewes enjoying the handsome creature and grazing together on the fresh green grass created by the rain. To my dismay they still huddled in the corner of the fence line with their heads down, and to my amazement the ram had joined them. They didn't eat all day, but by nightfall my husband had figured out what to do with them. He led them to the gate, opened it, and the flock scampered over the broken twigs and ran back to their place of safety. We left the gate open, and they ventured through at their own pace, accepting the new pasture within a week or so.

"O Lord, is that the way it is with me? When I am far away from Jesus, separated from the fold, I am so insecure and frightened that I don't eat properly, spiritually or physically, and my head hangs down in depression. Sometimes I even stand around in groups, complaining and murmuring.

"Praises to You, my Shepherd, who comes to find me and brings me back to the fold. Who allows me freedom to make choices, even if they are wrong, and leaves the gate of indecision open until I can find my way through to You." ♦

The Shelter

THE rain pelted on the tin roof as we struggled to get the pieces of roofing nailed down before another gust of wind sent them flying through the air. It was late Sunday afternoon and we had almost finished our lamb shelter. But the storm would not leave us alone, and our hands were wet and cold. Night had fallen. Flashlight in hand, we continued until we pounded in the last nail. At last we could relax—or could we?

Our ewes were ready to lamb, the roof was done, but the side walls of the shelter still had to go up. We had no time until the following Sunday to complete the project.

"O Lord," I prayed, "please keep our little flock safe. We aren't ready yet for their lambs."

The next Sunday found us out in weather about as stormy as the week before. My husband worked hard sawing and nailing boards. I followed behind with paint can and brush. That night we breathed a sigh of relief. The lambs could now be born. We were ready and so was the shelter.

It was interesting to follow the progression of the births. The ewe we had named Topaz, afraid of the shelter, had one twin out in the rain the following Saturday evening. We found her little ewe lamb and brought both mother and newborn into the soft straw bed. A little ram lamb was born within a half-hour. Just before dawn on Monday morning Tara had her twins outside in the grassy meadow, but she quickly adjusted to the shelter as a new home for her newborn ram lambs. Sweet Sixteen, our third ewe, lambed her twin ewes in the sunshine between storms right in the middle pasture—a perfect view from my kitchen window. By the time I arrived on the scene, I had nothing to do except help clean them with a large bath towel.

Our lambing pens were full of sweet smelling hay and huggable newborn lambs that Monday night. My last chore was to adjust the electric lamp, put fresh water in all the buckets, and give new alfalfa to each ewe. As I walked up to the house in the softly falling rain, I felt extremely thankful that my husband had spent many long hours constructing a door for easy access to the barn and had made individual pens for each ewe and her offspring. A good night's sleep was long overdue!

Several days later all the ewes were out grazing in the lush green pasture with their little lambs frolicking about in the wildflowers. Then the sun disappeared behind a dark thunder cloud, and I could see it would be storming again. Quickly I went out to put everyone in their shelter.

While I secured the last gate, the rain fell in uncontrollable torrents. As I stepped under the tin roof, I prayed:

"Lord, thank You for providing a shelter for us, the lambs of Your pasture, where there are fresh, sweet meals from Your Holy Word. The clear and sparkling Water of Life is so freely given. You see the storm approaching and You are out there directing and guiding all Your flock to come under the roof of Your protection. And You even provide a structure called Your church. Help us not to be afraid of that shelter. You are the true Shepherd who is guiding us through the open door. It doesn't matter to You that some of us have experienced new birth in the storm, rain, early dawn, or even in the bright sunshine of our lives. What is important is that we have been born again, and You have provided a place of safety for all Your lambs. 'Surely, goodness and mercy shall follow me all the days of my life, and I will dwell in the house of the Lord forever' " (Ps. 23:6). ♦

The Stickers

THE spring pasture grass had turned to summer gold. Weeds and stickers abounded. Each evening when we locked the sheep up for the night, we checked our small flock for stickers. We especially looked around the eyes, for once a sticker got inside the eye it was painful to remove them. Many times we would have to pull foxtails backwards from the soft, pink, tender eyelids, and I'd cringe as I removed them as quickly as possible and put an antibiotic ointment in the eye. I felt so sorry for our ewes and lambs.

"They have no hands with fingers to pick out the stickers," I bemoaned to my husband as we walked arm in arm up the path to the house. "They are totally helpless," I continued, "and they really do need a shepherd to care for them."

"Lord, that is the way it is with me. I need someone to help me with the beams and the motes. The splinter in my 'brother's' eye is so small and the plank in my own eye is so large that I can't possibly see to remove his. Only as Your fingers of love move over my soul with gentle pressure will the beam ever be removed, for I am as helpless as my lambs. I need You, my Shepherd, to anoint my head with the oil of Your Spirit to ease the healing process. Then, after You have anointed my head, with eyes now filled with love, I will no longer see the speck in my brother's eye, and if I do, I'll know how to help him remove it, with Your hand of love." ♦

Lambing

ANYONE who raises sheep knows that when lambing season begins, all other activities screech to a halt. Laundry piles up in the hampers and dishes stack up in the sink. That beautiful spring whirlwind called lambing interrupts everything.

We were busy that particular year, up night and day, helping the ewes in labor. Some were to have their first lambs, others were veterans of the event.

I was so occupied with lambing that I didn't have enough time left to drive to church for the meetings held by the visiting evangelists. When all the lambs had safely arrived, I accompanied my husband to church one evening, only to find that while the evangelist was in the main sanctuary, the cooking class met in one section of the church and another class convened somewhere else.

"O Lord, don't we, the sheep of your pasture, understand about 'lambing time' in the true sense? When the evangelist comes to town or the pastor holds a series of meetings, all normal activities should stop to allow full concentration on the task at hand. It is important to help those newborns come safely to us through the channel of baptism. Accidents can happen in the birthing process, and we need to be there with outstretched arms to assist them as they enter the new world of Jesus' love. Help us not to be so busy with mundane programs of church life, needed as they are, that we aren't willing to stay up night and day if necessary to witness the most exciting event: a soul being born again." ♦

New Barn

THE timbers were large, heavy, and thick, meant to support a pole barn. The trusses were up and the flooring down for the second story, making shade underneath. For months I watched this unmoving skeleton as I fed my sheep morning and evening in the west pasture. Many times the silhouette, as the sun set, reminded me of a beached whale. My new barn was unfinished, without roof or siding.

Then the heavy spring rains unleashed upon us. The pregnant ewes, wise enough to know the "barn" offered some relief from the sun, came running for protection from the heavy downpour and whistling winds. Night after night they came for shelter under the roofline, standing there bewildered with their situation. The rain leaked through with almost as much force underneath the second-story floor as it fell outside.

There stood my poor sheep—wet, cold, and miserable—waiting patiently for nature to bring on the proper time for their lambing. They looked at me as if to say, "If this is supposed to be a barn, why is it raining in here?"

"O Lord, are we like that sometimes? Does our church structure look large and impressive, providing an illusion of peace and safety from the storms of life? Are the sheep of Your pasture, the lambs of Your flock, seeking the protection of Your church in the sunshine of life, finding that when disaster pours down in a torrent of confusion and tragedy (physically or spiritually), they also experience the same bewilderment? We too are waiting, Lord, for the birth of that grand and glorious day of Your return. In the meantime, please help us patch up the spiritual leaks with the bonding cement of Your love, and have a strong solid roof of truth over our heads. And never let it be said by anyone 'If this is the church, why is it raining in here?' " ♦

Joy

THE pressures of the lambing season were doubly heavy that spring as I also had the full time care of my father. Growing frail, he needed a great deal of my time. The lambs were due anytime. Just thinking about my responsibilities left me exhausted.

Immediately after supper that late February evening, I walked down to the barn to check on the ewes. Spotting a beautiful white newborn lamb, I grabbed some towels and wiped him down, then ran up to the house to get extra towels and tell everyone that the lambs were arriving. By the time I returned to the barn, Rain had given birth to another ram lamb. We were pleased they were strong and vigorous newborns.

Two weeks later on a Tuesday afternoon out in the pasture Button delivered a set of beautiful black twins. However, they were premature and needed extra care. Sunday morning found me in her pen cradling a dead ewe lamb in my arms. I wept over the accidental death of one of Button's preemies who had already grown healthy and strong.

Two weeks later we were out on the pond dam helping a ewe who was having a difficult labor. After the birth of the first lamb, we had gone to make a phone call for assistance and returned to find the ewe had triplets—one stillborn, a ewe lamb. I felt discouraged and depressed.

"O Lord," I wept, "I have never had a year like this before. Please help us through the rest of the lambing." So far all my ewe lambs had died, and for a breeder, it was a real loss.

Sunshine had a set of twins on Friday evening: a ram and a ewe lamb. At last I had a little ewe lamb. Sabbath morning Tara was in labor as we left for church. We expected to see a baby at her side when we returned. By early afternoon I realized she was in trouble, too, and called a friend to help. Two beautiful large black babies were finally delivered from their mother: a ewe and a ram. Then we pulled out a surprise: a white, weak, under-developed ewe lamb came sliding out with our aid. Her feet were crippled and her mother rejected her. "I have no time or energy for this," was my mental prayer.

While I didn't want to feed a bottle lamb every four hours, I did still have three ewe lambs, and chose to accept the challenge of raising an orphan animal. It was easy to name the three lambs Faith,

Hope, and my little "bummer" lamb, Charity. She truly was a love, and her name fit her more each day.

The following Tuesday morning it was all over. The vet had just left, our last set of twins had just been delivered, and I breathed a sigh of relief. The little ewe lamb was black and so very beautiful. On impulse I named her Joy!

Walking up to the house, I was glad lambing season was finished. Stopping by my father's bed to share the early morning events, I let him know all was well. He looked up and asked, "Remember what David said?"

"No, dad, what did he say?"

" '. . .weeping may endure for a night, but *joy* cometh in the morning' (Psalm 30:5)."

I kissed his cheek and smiled, wondering how he knew I had named her Joy. ♦

Charity

AT her birth, I didn't know how true her name would become. Her arrival in third place in a series of three ewe lambs born in two days almost predestined her. But Charity's name seemed too large for her at first. As the days passed, I decided this weak, sickly ewe lamb with crippled feet didn't have an appropriate name. Maybe I should reshuffle the names, calling her Hope, because I seemed to be constantly hoping and praying that she would live. But the little white lamb had a tremendous will to survive. The creature's nursing siblings could not match her sucking instinct. She wanted to live!

Her mother's rejection of her at birth gave me new opportunities to become more skilled as a shepherdess and stretched my commitment of raising sheep to the limits as I continued to care for her. By two weeks of age her feet had become normal, but she had pneumonia from milk flowing into her lungs as she sucked with vigor on a nippled Pepsi bottle. Then she became bloated and had to stay at the vet clinic for several days on intravenous feeding. While she was there, the vet discovered she had the parasite Coccidia which causes gastro-intestinal problems. Medication and other complications created continued stomach upset which, in turn, caused more digestive problems. Then she suffered from a fever and a return bout of pneumonia.

I had no doubt in my mind now—Charity was indeed her name. She had a pleasant disposition and her mouth turned up at the edges, creating a constant smile. Long-suffering, she had kind eyes and the spirit of a survivor. My most affectionate lamb, she'd stand at my feet, snuggling close to my legs for protection, not even aware that I was not her real mother. The small creature trusted me. How could I let her down? As I gazed out over the sheep in my pasture, there I saw abiding in my field, Faith, Hope, and Charity . . . and the greatest of them was Charity.

"O Lord, please may these attributes that I have witnessed so beautifully in this little innocent lamb

be seen in me. She has given me anew the aware-
ness of Your tender care for the flocks of Your
pasture. As the Good Shepherd in my life, please
heal me from parasites and other ills. In my eager
desire for food from Your outstretched hand, may
I not choke on Your Word as I seek to survive on
a daily basis. Father, I am looking up to You from
the foot of the Cross. I trust You for I see Your Son
the Lamb of God who was rejected, dying that I
might live. Please forgive me for causing You pain.
As You continue to heal, may You see faith, hope,
and love abounding in the pastures of my life. But,
may the greatest of these be LOVE." ♦

Twins

FIFTEEN hours!" I exclaimed. "That's impossible.
No newborn lambs could live that long without
colostrum." I clutched the telephone in excitement.
But it was true. They were still alive and Candice
had taken the twins to her house to feed them
frozen colostrum (the first milk secreted after
delivery) and encourage the modern-day miracles
to survive.

Revulsion swept over me as I put the receiver
down and occupied myself with daily chores. "Anyone who would allow a ewe to breed without any
bag or nipples for nursing must be a fiend," I
concluded.

This man, who raised sheep several farms
away, had made some kind of bargain with Candice, knowing she was a veterinarian's assistant

and had an excellent knowledge of sheep, to come and pick up the lambs when they were born. He had neglected to butcher the ewe the year before when she had lost her bag to mastitis.

Candice had explained to me on the phone that the twins had been born at three o'clock the previous morning and no one had called to let her know of their birth until six o'clock in the evening of the next day. She had given explicit instructions that whenever the ewe lambed, the farmer should call Candice—who would come any time, night or day, to pick up the newborns.

The first thing a newborn swallows from its mother is colostrum, which provides strength, antibodies, vitamins, and protein. This nourishment provides the lambs with a better chance for survival. The farmer had not only neglected to put the ewe down, but had also not bothered to keep her from his ram.

"All those who raise sheep are not always shepherds," I thought as I sorted out my angry emotions.

I recalled the last part of our phone conversation. Candice had mentioned the hardest part of the journey to pick up the twins was when she had put them in the back of her camper. The mother ewe was extremely attentive, trying to be a good mother, yet totally unable to perform her duty. She dashed alongside the truck and cried wildly for the babies she would never see again. Her owner was now ready to butcher her.

My thoughts crystallized. "O Lord, I am so thankful that in our maternal hearts You not only put the milk of kindness, but provide ways for us to feed and nourish our children as they grow. You promise, 'Train up a child in the way that he should go; and when he is old, he will not depart from it' (Proverbs 22:6). And You also enable us to do what You have asked. Truly You are the Good Shepherd who gives all Your newborns the precious milk of Your Holy Word to nourish us in Your bosom after our new birth experience.

"We praise You also that You are not a butcher, a fiend, letting us run unattended in the fields of life. You are there in all of our activities. And moment by moment You are with us through Your Spirit that the Bible writer speaks of as 'that great Shepherd of the sheep, [to] equip you with everything good for doing his will, and may he work in us what is pleasing to him, through Jesus Christ . . .' (Hebrews 13:20, 21). And last but not least, Lord, may all those who tend your flocks be not just producers of sheep, but real shepherds, is my prayer." ♦

Time

WITH addresses and map in hand, Candice climbed into the navigator's side of the pickup. I drove. We were out to find the "ultimate ram." Every early autumn black-wool sheep farmers go through this ritual.

We continue our search year after year because one never finds the elusive creature. Finally we all settle for what is available, or our old ram breaks down the fence to get in with the ewes and nature proceeds as expected. Lovely lambs are born in the spring despite the unfound perfect ram. However, breeders of small flocks like to rotate their breeding ram every two to three years, and it was my year to purchase a young one.

The two of us had decided to visit several farms within a 20-mile radius. We met one owner, looked at her ram, and drove on to the next ranch. The Natural Colored Wool Growers Association, a national organization of which I am a member, encourages the up-grading of colored wool breeds and had provided the list of addresses.

As we went from farm to farm the thought struck me, "These are lovely people with beautiful, healthy wool sheep, and none of us knows the other exists." The book listed our names in alphabetical order, not by towns or counties.

Penny invited Candice and me into her cozy country kitchen. Having completed the tour of the ram and ewes out in her pasture, we now sat on antique chairs around an old oak table discussing our yearly plight. Suddenly an idea occurred to us. Why not start a local group of natural colored wool breeders right in our area. Penny became as excited as we were about the idea. All kinds of creative ideas tumbled from our minds. It could be a support group during lambing or other stressful times in our sheep business. We could encourage each other in the sale of our fleeces and even have a ram exchange in the fall. With about 17 farm families within calling distance, there was no need to drive from farm to farm each year. The other two women assigned me to make the phone contacts to organize the group.

The enthusiasm on the other end of the telephone line was exhilarating. I talked for over half an hour with each person I contacted. Each spoke expressively about black sheep—the length and silky texture of the wools and the beautiful shades of deep charcoal, gray, brown, black, and muted earth tones mixed within individual fleeces. They discussed lambing time, shearing time, and breeding time. Once started, they were hard to stop. Everyone was eager to share with me their delight and love for their flock. When I put down the receiver after the last conversation, I was amazed.

I'd made phone calls before to share with our church parish members about an afternoon potluck, or to encourage them to attend Week of Prayer readings in a certain member's home, or to attend a vespers featuring a guest speaker. The response had been so different.

"Could it be, Lord, that these new sheep friends spend much time with their sheep, while we, the sheep of Your pasture, spend such little time with the Shepherd that we have nothing to say?" ♦

Gee Kee

THE early morning caller was happy that she had

found a home for her pet ewe, a bottle baby given to her and her husband several years before by their granddaughter. Now unable to care for the animal because of ill health, she telephoned our home. She was pleased to know we did not eat our sheep and felt safe in making the offer of a free Cheviot-Barbados cross bred 4-year-old ewe.

"My reputation for taking in unwanted and discarded pets must have preceded me," I mused as I hung up. I thought she knew that I raised black sheep, so I neglected to give the color of the ewe a second thought.

Kathy, my neighbor, and I got into her truck with the attached horse trailer. The journey was longer than we expected, and by the time we drove up the circular drive to the wooded country home in the higher foothills of the Sierra Nevada, I was glad we had found the place at all.

Disappointment struck me when I saw the ugly white ewe in the side pasture. Had I traveled all that distance for this? Besides, I had given Kathy gas money for the use of her truck and trailer. Too embarrassed to say anything, we loaded the animal and drove home in silence. I didn't even like the ewe's name: Gee Kee.

Raised on a bottle and handled by humans, Gee Kee only related to people. The pasture where I placed her had six other ewes and she would not eat or drink with them. Otherwise a loner, she would approach the fence and bleat in an unpleasant, raspy, irritating call for attention any time someone went out the back door. She was always there, eager to be petted. Her beautiful eyes were her one redeeming feature, and all who visited us admired her unique personality. My husband enjoyed her more than any of our other ewes because she loved to have him scratch her ears. As long as anyone would scratch them or rub between her front legs (which she also enjoyed) she would stand there soaking up the attention.

Gee Kee had never been bred, and when the ram arrived on the property in the fall of that year, she instantly accepted his attention. Where the ram was, she was also, grazing right beside him. It was fun to see her adjust to her new-found role as a sheep.

When she lambed in the spring, I finally fell in love with her. With the help of the veterinarian, she delivered the biggest lambs we ever had. Unfortunately, though she had earned a real place in all our hearts, she could not adjust to being one of the flock. She kept her offspring close to her and away from the other sheep. Except for humans, she knocked, butted, and crashed into anything that came near her or her babies. Her aggressive behavior became a threat to the rest of the herd. After the lambs were weaned, I decided to send her off to auction. It was a hard choice because I was emotionally attached to her. However, for the safety of the rest of the flock, ewes and lambs alike, she had to go. As I helped load her in the vehicle that would take her to the auction and her ultimate death, I was unable to shake the thought that no one there would save her.

My mind whirled. "Lord, is that the way it was with me? I've been a loner, standing proudly by myself with my head up high, not like the other sheep in Your pasture. As a result I've not had much tolerance for others who got in the way, old or young. So recalcitrant, unable to change. When circumstances drove me to the auction block of life, You were there."

I remembered the prayer I wrote so many years ago:

Freedom

THE auctioneer raised his voice
To an even higher pitch.
Here, this slave will work for you
And do your every wish.
For I had sunk so deep
My mind was so beclouded
Truly was I bound in sin
In death robes was enshrouded.
But the Shepherd came to bid for me
And for my heart to seek.
Didn't He fully realize
I was lonely, sick, and weak.
Upon the auction block I stood
In pain and shameful loss.
I saw Him pay my price
By dying on the cross.
His head bent low in agony
Naked, in deep humiliation.
He paid the price for each of us
All mankind and every nation.

When He came to purchase me
I was undone and bare.
He provided a robe so white
Of righteousness to wear.
"I bought you child, to set you free
This was my true endeavor."
Now at his feet I humbly fall,
I'll serve you Lord forever.

AMEN

Hit by the Ram

THE anticipation of a week's vacation at Lake Tahoe and attendance at camp meeting while there got me off schedule that warm August morning. It was Sunday, and we had just finished eating breakfast.

We were still sitting with our elbows on our round oak table discussing farm chores. Chris, our son, and Laurie, our daughter-in-law, would watch the place while we were away. It would be our first trip since my back surgery in February. The warm breeze wafted through the open windows and reminded me that I had better run and take care of the rams before I forgot again.

The rams were frustrated at not being fed on time. As I approached with the wheel-barrow I could sense their frustration. By now they were pacing back and forth along the fence line. I pushed the wheelbarrow of alfalfa ahead of me as I entered the dry pasture that the two rams shared with two wethered goats. When I threw some feed toward the far side of the wheel-barrow, the goats ran toward it while the two rams charged toward me. While the mass of alfalfa was still in midair one of the rams knocked me down by ramming my left thigh. Although the blow did not break the bone, it did cause my knee to go out of joint. Concerned about my back, I grabbed the fence behind me and slowly sank to the ground. The rams meanwhile had seen where the alfalfa had fallen and were now mildly munching on it.

"Help, I have been hit by the ram, help I have been hit by the ram!" I shouted over and over.

The next thing I saw was my husband and son bursting out the kitchen door at the same time, screen door banging behind them. Chris was carrying a shotgun and Chuck was running to grab the pitchfork that leaned against the tree. They ran down to where I lay in the pasture.

All I could think of was, Why are they going to kill the ram? He didn't mean to knock me down. They were just excited by hunger, and when the goats got fed first, they started for the wheelbarrow. Unfortunately I happened to be between them and their food. It was an accident. So why would they want to kill the ram? "Don't kill the ram, don't shoot, he didn't mean to hit me!" I shouted.

When the men reached where I was sitting up against the fencepost, they demanded in unison, "Where's the rattler, where's the rattler?"

Staring at them in astonishment, I said, "I've been hit by the ram!"

Everyone laughed as we realized how silly we all looked. Me on the ground, Chuck with a pitchfork, and Chris with a gun. They had heard my voice through the open windows saying, "Help, I've been bit by a rattler!"

Laurie had gone to dial emergency 911 for information on snakebite and the men had raced out to kill the snake.

The two men loaded me into the wheelbarrow and hauled me back to the house. My knee was badly wrenched but we had a brace from an old ski injury that would hold my knee in place. They finished loading the car, including me into it, and we drove off for our vacation, brace and all.

As we traveled over the miles I thought about the events of the morning and the bizarre twist of what everyone thought they had heard in the house. Glancing over at Chuck, I asked, "I wonder if that is how gossip gets started?"

He looked at me and we both smiled. ♦

GARDEN PRAYERS

"Though the fig tree does not bud and there are no grapes on the vines, though the olive crop fails and the fields produce no food, though there are no sheep in the pen and no cattle in the stalls, yet I will rejoice in the Lord, I will be joyful in God my Saviour" (Habakkuk 3:17, 18).

Garden of My Heart

LORD my heart was a fruitless place
Where briars and thistles once grew
With patience You've tended and pruned it.
I give love and praises to You.

The pruning so painful that I cried
I see now through the heartache and tears
The rain from heaven never stopped
Softening my hard heart over all these years.

As You watered my heart with Your spirit of
    love
My face was turned toward Thee.
May the fruit that I bear show forth to all
That Jesus is living in me.

You've watered the garden of my heart
With Your spirit sent from above
Thank You dear Jesus for caring so much
And sending me showers of love.

The fruits of the spirit
Will nevermore cease
They'll bring forth a bounty
Of love, joy and peace. ♦

Transplanted

IT was just a little seedling
That I found there one hot day,
Not amongst the other vines
To catch the garden spray.

The vine was almost dead
It must be transplanted soon,
I must do it quickly now
For it would die by noon.

So, I began to water
In its new and fertile place
And maybe one day soon
Tomatoes I would taste.

And now the plant has grown
And there's tomatoes on the vine
And this precious little plant
Is special 'cause it's mine.

I've related this to you
So you could clearly see,
For in mercy He saw me dying
And this is what He did for me.

He gently had to transplant me
He pulled me roots and all,
He placed me in a fertile place
So I could hear His call.

He tilled the soil around me
And a new heart put within,
Now I know why He cared so much
For I belonged to Him. ♦

God's Other Book

TAKE time to sit and meditate
Upon my handiwork of earth,
To smell the flowers on the path
Springtime blossoms, the joy of birth.

My child, please notice in the sky
The moon and setting sun
For you I made these brilliant lights
Before creation week was done.

I long to tell you of My love
The brook and rolling sea,
All nature speaks with words so clear
Please spend some time with Me. ♦

Helen

HELEN was a beautiful neighbor. She had warm blue eyes, strawberry blonde hair, a smile that radiated gentleness, and a sense of humor. I loved to hear her laugh.

We had first met over a decade ago when I rattled her fence gate after having noticed her hand-painted sign: "Brown Eggs for Sale." She lived at the beginning of our one-mile, dead-end country road, and I resided near the end. I passed her orchard on my morning jogs and watched the seasons go by from blossoms to spring green to fruit to the falling of the leaves to winter.

Helen was my summer friend. When I would purchase her sweet red raspberries, cherries, lemons, grapes, and brown eggs, she'd give me seedlings for melons, squash, tomatoes, artichoke plants, and anything else she had in abundance. We talked for hours as she helped me harvest my purchases. I envied her cornfield. It always stood lush with promise, fertilized with the best the chickens had to offer.

As we picked crops, I discovered that she and her husband had retired early to enjoy the country life. Her daughter and family lived next door on an adjoining parcel of land, with a garden gate between so the children could visit their grandmother anytime.

One of the last times I talked with her was in the late summer. She had broken her thumb and my husband had cut part of his finger off in a power saw accident. They had common pain and she expressed her sympathy for his injury. But time flew by with the rush of fall activities, Thanksgiving, and Christmas, and I lost touch with her.

Finally the quiet of a new year came and gave me time to think again. I hadn't seen Helen for a number of months. On the way home from a shopping excursion, I saw her new next-door neighbor walking her dog and children. As our eyes met I stopped the car.

"Did you know Helen died last week?" she said hesitantly.

I wanted to scream, "Why hadn't someone told me?" Instead I silently protested, "No! That can't be possible. She is my friend, my summer friend. She's been there every year since I moved out from town."

Helen had taught me so much about gardens, had helped me take a sick goat to the vet, and had shared the bounty of her labor from her garden and orchard. And now my summer friend had died in the winter!

"O God," I wept, "why was I not there when she needed me in the off-season? Please forgive me! O, make me a neighbor for all seasons, in the dreary winter as well as the summer of life. Let no time pass without me being aware that a friend is missing. For I hear You gently whisper, 'Preach the word. Be ready in season and out of season . . .' (2 Timothy 4:2)." ♦

INDOOR PASTURES

"My sheep listen to My voice. I know them and they follow Me. I will give them eternal life, and they shall never perish; no one can snatch them out of My hand"
(John 10:27, 28).

The Curtains

SEVERAL seasons went by before we got around to selecting curtains for our country home. Finances were a factor in choosing just the right window coverings to enhance the beauty of the "early attic" decorating theme. For a few short weeks my daughter would be visiting, so I called my mother to ask her to help. We would all meet at my place for a sewing bee.

Our goal was to purchase fabric, sew, and hang in place double Priscillas for ten large windows set in the walls of our living room and dining room areas. We took on the task with gusto, little realizing the challenge that awaited us. Since we were all seamstresses of some merit, we expected to complete the job within our limited time frame.

Once we had selected the fabric; we turned the dining room into a miniature curtain factory, measuring, cutting, pinning, basting, ruffling, sewing, and ironing the sheer off-white organdy. But confusion abounded, and we found ourselves wasting time and retracing our steps.

Stopping, we looked at each other and laughed. Somehow we had to bring order out of chaos. My daughter Karen was best suited to stay put on the ruffling machine. Mom would sew straight seams as rapidly as she could, while I was assigned first to the floor to cut the fabric in even strips for curtain length and ruffles. Then on I'd go to the ironing board when each curtain was complete. Because we had an assembly line, our job worked smoothly and we finished on schedule. Each of us felt great joy as we each found our niche and together created a thing of beauty. My windows now are framed with the most professional-looking, soft, finished curtains, creating a lovely glow when the sun streams through the panes morning after morning.

"Lord, is that the way it is with me when I want to work for You in the church structure? Eager to make something beautiful, I rush here and there. I try to do so many jobs that sometimes responsibilities don't get completed on time. May there be no confusion in the role You have assigned me, even if it is on the 'threshing floor' of life, for there I see the chaff removed and the grain separated out for service. Help me accept my niche so together, we as Your church, can create a thing of beauty, and the Son will glow through each individual day by day." ♦

Polishing

As I sit here in the cozy warmth of my family room, enjoying the crackling fire, I can see how the kitchen floor glistens with the rich dark tones of fresh wax and a thorough polishing. It even smells good.

Somehow the beauty is marred, however, in the knowledge of how it was completed. The carpet man had come to clean the carpets, and I had to do the wood floor adjoining to put it all in a clean balance. But the sun set before I had finished. The sun sets many evenings before I complete a task at home, but this sunset marked the beginning of those precious 24 hours of God's Holy Sabbath. I felt bad, for that Friday nothing seemed to happen in the proper order. The phone rang at the most inconvenient times as I rushed feverishly to polish the last portion of oak flooring, but the Sabbath began before I completed the job. The preparation day had ended and I wasn't ready!

"O Lord, will it be like that for me in that final hour, hurrying about to put everything in proper order and balance, polishing here and there on my character? You have given me a weekly cycle to prepare for Your Sabbath week after week, and a preparation day, a symbol in part to show us the hour in which we live. Please forgive me for even thinking that I could ready myself. Only as Your Spirit infuses me will I ever be prepared. Mold this earthen vessel into a shape fit for Your use. Please don't cast it aside as worthless because of my disobedience. Forgive and fashion me anew with the water of your Holy Spirit. Soften and subdue me and turn my eyes heavenward, for on the horizon of time I hear the sound of Your coming. Our years of preparation have almost ended. Even so, come quickly, Lord." ♦

Spinning Wheel

IT was time to purchase my spinning wheel. I wanted an old-fashioned one with beautiful hand-carved oak wood, solid, polished, and perfect. It would be a wheel to admire, an antique to place in the corner of my family room, shouting for all to see that a spinner lived in the house. It would make me feel like a professional and not the novice I was. Just to see it daily in the room would be an inspiration. But I settled for one of the new modern varieties: functional, plain, and affordable.

After my daughter assembled it, the wheel was so small no one ever noticed it in the room. You could see the skeins of yarn displayed artistically, the natural colored wool in baskets with their warm tones, several knitted scarves and hats, and wool in bats and roving (twisted strands of wool) ready for my soon-to-be-deft hands. But the wheel stood unnoticed and unused.

As the days passed I was so busy that I rarely had time for the spinning wheel my husband had purchased as a gift of love for me. Several of my spinning friends came by to encourage me and explain how they were improving their skills by spinning a certain amount each day. I thought about it often on my morning jogs.

"Lord, I have spun some beautiful lovely yarn, and made a few scarves and hats for friends and family. Now I am thinking of buying a new type-writer, or even a computer, to enable me to spin more quickly the threads of truth from the treasured gift of writing You have given me, and knit them into garments for the soul. Let it not be something impressive in the corner of the room where I write, so people will walk by and say, 'You must be doing a lot of writing now that you have this new modern technical tool.'

"I have seen already how I have not used the potential of my spinning wheel. I am concerned about my new, yet-unpurchased computer. Will all the modern conveniences make me sit down and do the work, spend the time needed on a daily basis, using my moments wisely?

"O Lord, is it old fashioned discipline I need? I see in the difficult word 'discipline' the beautiful word 'disciple.' Is that what You want to do with me? Discipline me today for the disciple of tomorrow?" ♦

Heavy Boughs

THE festivities had arrived and gone. Gifts opened, hugs and kisses exchanged by family and friends, meals devoured, cookies and fresh fruit eaten, a brand new year had begun! As I took down the tree lights and festive candles, removed wilted poinsettias and all the other trimmings, placing them in boxes for storage in the attic, I wondered about the gifts Jesus so bountifully bestows on us—His life and the beautiful fruits and the gifts of the Holy Spirit. Have they languished unopened under the tree of my life, like the small wrapped package I found last year beneath the heavy, decoration-laden boughs. It had gotten lost under all the tinsel and festive lights. Or would I place Jesus in a neat little box someplace in the attic of my mind, waiting for the next year to bring Him out again to dust off and display?

"O Lord, I plead, deliver me as I seek to put my life back in order for this new year. Help me with my deep commitments and resolutions. Trim and lift the heavy-laden boughs of my life, take away the tinsel, and let me find any gifts I've overlooked.

May the basket of Your spiritual fruit which includes love, joy, peace, and long-suffering, be found in my life, and may I use the spiritual gifts I already have for unity in the body of Christ. And, Lord, may I never display You on the mantel of my life. Please, I pray, live out Your life within me." ♦

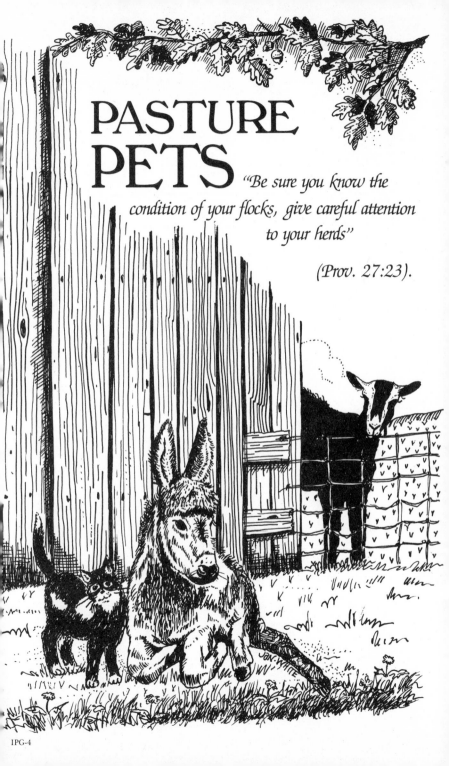

PASTURE PETS

"Be sure you know the condition of your flocks, give careful attention to your herds"

(Prov. 27:23).

Mother Goose

SHE came with expectancy in every movement, wanting to see her pet goose and the two downy soft yellow and gray goslings, which had been left several days before on our farm by her husband, who had moved out of their broken lives. I sought to reassure her that the geese would be safe and we would provide a good home for her much-loved pets. Her eyes and those of her children revealed the loss of a family unit and all their farm friends as well. I hoped to ease some of their pain. We walked to a depression in the ground that had filled in with water. It was a makeshift arrangement, to be sure, but the goose and her babies were splashing playfully in the water. The happy scene abated some of the broken family's anxiety.

"It has been so hard to find good homes for all our pets, and we still have a cow and two horses to relocate." She spoke with a distant look in her sad eyes as she and her family returned to the car.

The weeks passed quickly and the goslings grew wings and feathered out to be beautiful gray and white birds. We had come to love them. Their antics in the pond reminded me of amusing clowns. They loved to show off when an audience appeared along the pasture fence. Mother Goose and her babies were doing well. By early spring, her young ganders had grown bigger than their mother.

One stormy Sabbath afternoon, we returned home from church to find a stray dog in the back pasture slinking out of the hen house. The quietness that greeted us alarmed me, for the geese usually made loud "welcome home" noises. My eyes quickly surveyed the pasture—no hens and no geese! I ran out the back door with rain coat flapping in the wind, rushing toward the gate. The hens were safe in their house. But my geese, my beautiful geese! From them came a terrible stillness.

I found Mother Goose in the far corner of the pasture where she had died trying to protect her young, seeking to fight off the predator with her lovely gray wings. Also dead were the two young ganders, one in our field, another drug off to an adjoining pasture. A neighbor dog had gone mad. It was a senseless waste, and one I could not change.

"O Lord, I am so glad that you can change things, that you can mend broken families, that you provide a place of shelter, that you defend us with Your wings! As long as we choose to stay under their protection, You have power to save, and there will be no senseless waste as the mad dogs of this world rush to destroy us. And on Your Holy hill and during Your great millennial Sabbath and throughout eternity, You promise there will be no more death, for the wolf and the lamb will feed together (Isa. 65:25). And in our time of trouble, You will cover us with Your feathers, and under Your wings we will find refuge, and Your faithfulness will be our shield and rampart (Ps. 91:4)." ◆

Dry Pasture

THE pond was beautiful, its edges trimmed with short spring-green grass. The reflection of clouds and sky created moving pictures on the smooth still water. Sometimes it was difficult to see where reality ended and reflection began, the colors of trees and sky blending so perfectly. My new geese loved the pond. Swimming, gliding, splashing, and flapping their wings, they played all their goose games, enjoying every moment. Oh, if spring would only last forever. But summer came. Slowly the waters receded until the last tadpole perished and only a dry, crusty, barren wasteland remained, an empty hole in our dry pasture.

Each day I'd set out buckets of fresh water for the geese to drink and play in, and fill the barnyard bath tub to overflowing with extra water. Unfortunately their beautiful white feathers became dirty and lifeless. They molted heavily and scattered their feathers everywhere. My poor geese were just surviving from day to day. Wishing our water system was adequate for irrigation, I knew that until the fall rains began again, I'd just have to provide the best I could.

One morning as the water dripped from the opening in the front part of the old tub, it formed a little stream and flowed into a dip in the ground. On their daily journey to watch me fill that tub, the geese stopped to bathe and splash in the little puddle.

"Lord, is this what you want of us? To come daily and be refreshed by the water of Your Holy

Spirit, like the widow in the Bible who had just enough oil for her daily needs? Someday the latter rain will fall, but in the meantime, may I come daily to be revived by the cool clear water of Your Spirit that You faithfully offer me. In my own dry pasturelands may I go about the duties at hand this day, filled and satisfied with my daily portion of the Water of Life, so freely given." ◆

Pretty Goose

THE day before Christmas we opened the gate to allow my neighbor, Pat, to bring her three horses, pony, and two young heifers to graze on my west pasture. My four new white Chinese geese were floating across the pond like ships gliding smoothly on an imaginary journey. They made a lovely upside down reflection on the glassy water.

"What a beautiful day," I thought as I gazed out the kitchen window.

Suddenly, without warning, Pat's horses stampeded from the meadow, racing past the pond and up towards the middle fence line. Panicking, the geese flapped their long wings wildly, propelling themselves out of the water and into the path of the thundering hooves. I raced to the kitchen door and screamed at the horses as I ran out, but the damage had been done.

Fortunately, only one goose was hurt. One I called Pretty Goose had a broken leg.

The accident occurred before Pat had even reached her home and she came running back. "Is the goose badly damaged?" With disappointment, she saw the wounded fowl trying to struggle to her feet.

We picked up the goose and rushed her to the vet. After the exam and much discussion, we decided to have Pretty Goose's leg placed in a splint.

The accident couldn't have occurred at a worse time. Pat had company coming for the next few days. My house bulged with family from near and far for the Christmas holidays. We had no time for such extra added responsibility. Recalling that many people eat the birds for Christmas dinner anyway, I wondered why I was trying so hard to save her.

Reluctantly, I brought Pretty Goose home that early Christmas eve, saying to the Lord, "I don't have time for this. If You want her healed and made well again, You'll have to do it."

All during the next few days the vet's instructions rang loudly in my ears. "She'll have to have hot and cold treatments twice a day and her foot rubbed to keep the swelling down." But I was too busy to heed the admonition.

Several days later, I begrudgingly put her foot in the hot and cold water, for it had swollen by then.

"O Lord, forgive me," I prayed as I administered the belated treatment. I had not meant to cause her pain, and she really had a will to survive. Each morning she would raise up on one wing and opposite foot to greet me as I approached her cage. She had been trying and I had not!

As I recall the months of Pretty Goose's convalescence—watching her gain strength, learn to walk again, and finally swim with her family—I am thankful that I took the needed time to help her recover, that my prayers were answered, and that God is concerned when a "sparrow falls." If He cares that much about my goose, think how much He cares about each one of us. ♦

Max

IT was five o'clock on Friday evening. Chuck had just left with Max in the front seat of the car. Max was our family dog of 17 years. A Weimaraner, our beloved pet who was being taken to the vet's to be put to sleep.

I rushed upstairs with tears in my heart, hoping a bath would wash away the events of the day. No sooner had I submerged in the bubbles than I realized I'd not said goodbye to Max during the rush of putting him into the car. He had been a large part of our living for all those years, and I loved him dearly.

We bought Max as a six-week-old puppy on a rainy Sunday afternoon in January. When the children had wanted something to do to fill their house-bound hours, I had suggested hunting for a puppy in the want ads of the local newspaper. All of us gathered around the paper spread out on the family room floor, marking in red ink the phone numbers of any ad that seemed interesting. First we decided to compile a list of types of dogs, from

A-Z, one suited for each member of the family. Then we would visit all puppies, make a democratic vote, and the one with the most votes would come to live at our house.

As we tumbled into the car it became evident by looking at the map that the first puppies closest to our home were Weimaraners, definitely not in the correct alphabetical order. My original intention was to start with Airedales and work our way through the list. But these pups lived only six blocks away and it was raining hard.

We entered the warm kitchen wet and eager for a view of the pups in the large whelping box in the middle of the floor. They overwhelmed the children and me with their silver-gray, almost-nude, fat little bodies. The little creatures had the most beautiful shade of blue eyes I'd ever seen.

The owner quickly mentioned one pup who would have to be destroyed because of a small birth defect. The hair on the back of his neck and upward to the back of his head grew in a cowlick in the wrong direction. All eyes turned toward the puppy and then to my husband. With a whimsical smile he pulled his wallet from his pocket and said, "We'll take him!" That was the day Max came to live with us.

And today was the day he would die. I rushed from my tub with towel grasped firmly around me and struggled with the phone dial. "I must call the vet and have them wait," I told myself. "I didn't say goodbye to him."

By the time I finished talking to the receptionist, the tears started to flow. Max was already dead. Chuck would be bringing him home to bury him in the freshly dug grave in our backyard beneath the oak trees.

"O Lord, I didn't want to make that decision. I so wanted You just to let him fall asleep at the most convenient time." He had become so feeble and frail, his muscles wasted away. Almost blind and deaf, he could barely hear us call his name. Helplessly I had watched him wither away day after day. The kindest thing was to let him rest. But it was hard to say, "This is the day."

It had been a busy day of cleaning and Max always seemed under foot. He had lost control of his body functions and I knew it was time. Although I did not want to make the decision, I felt that I had no choice now. I had called Chuck to come home early from the office so we could have Max buried by sunset.

Why did I feel so sad? My emotions flashed through my mind. I had been angry with Max when he soiled my freshly cleaned carpet. Now I realized that I had made that phone call to my husband more in anger than in love and concern for my pet's actual needs. Max would have understood in his fashion, if I'd only been able to ask for forgiveness.

When we wrapped him gently in his blanket for burial, I reached out and patted his lifeless form. "Max, I'm sorry. I loved you so. You were the most gentle dog we ever had, and I just wanted to say goodbye." Why, oh why, didn't I say those words while he was alive?

"O Lord, I have an ailing father who is frail of body, his muscles withered away. The ravages of time have taken the life force and twisted and bent him into something barely recognizable. Please help me hold his hand and be able to say, 'O Daddy, I love you so. You were the best dad I could have ever had.' And when the time comes for him to rest may all my goodbyes have been said in love and not in anger, is my fervent prayer."

"And God shall wipe away all tears from their eyes; and there shall be no more death, neither sorrow, nor crying, neither shall there be any more pain: for the former things are passed away. And he that sat upon the throne said, Behold I make all things new . . . " (Rev. 21:4, 5). ♦

Thomas

A RUGGED fellow, he had a broad chest, a large head, and green-blue eyes that looked at me with a mixture of fear and a desire to get acquainted. A roving, abandoned black and white male cat that came to our place several years ago, he received the name Thomas because it seemed to fit. He definitely was a barn cat. No indoor life for him, no choice morsels of tuna, just dry cat food and a bowl of water to keep him going. And he could catch mice and gophers. Surely he could earn his keep. So Thomas got to stay.

As the months passed, he became an occasional pal. When not roaming, he would walk with me out in the back pasture as I went to retrieve a stubborn goat or mend a fence. He was special to me even if his square face was scarred and one ear partly disfigured from many fights. When I picked him up and held him tightly in my arms, he meowed and snuggled close—for he seemed to be afraid of falling. In a strange way I was attracted to this offbeat animal.

One day as I looked at him lying there, I realized that he was injured. My mind quickly scanned all the possibilities: cat fight? dog mauling? coyotes? hit by a car? As I started toward him, his pleading eyes met mine.

Somehow I had to help him. I didn't treat any of my other animals as I had Thomas. He had been an unasked for burden, an animal left by a moving neighbor, an animal I had not wanted.

"If he survives here, it won't be because of any help from me," had been my original feeling. Now his pathetic condition touched my heart. By the time Thomas and I reached the vet's, however, I had fully decided to have him put to sleep. He was sick and badly injured.

"Do you have any relationship with him?" Dr. Brown asked gently as I explained that Thomas would probably bite him during the examination. I nodded reluctantly, remembering how the cat loved to be held tight.

"Well, let's give him a second chance, then," the vet continued. Was it right for me to put him to sleep because he was wounded and I did not want to pay the price of giving him that chance, either in time or money? After we discussed and settled on the fee, we decided to grant Thomas another one of his nine lives.

It is amazing to see how he has healed. Even his deformed ear looks better. He has shown his appreciation for being granted another chance by staying close to home. Thomas follows me wherever I go, rain or shine. The cat is right there reaching up my pant-covered legs with his large front paws. "Please hold me tight," he meows.

"O Lord, today as I look up toward You, please hold me tight and don't let me fall. Thank You for paying the price and giving me a second chance when I was wounded and broken. Do not let me wander far from home, and may I follow wherever You lead in the rain and sunshine of my life. And when You come in angel clouds, gather all Your children tightly in Your arms and never let us go." ♦

Jubilee

THE empty horse trailer swung behind the pickup as we completed the turn into the driveway. My neighbor, Jan, had accompanied me to Penn Valley that summer morning to adopt a baby burro. The burros had been rescued from a Naval landing field near China Lake in Southern California. The other alternative for them was execution.

I had never touched a baby burro before except at a petting zoo. Now I saw about one hundred burros in all sizes and shapes, milling about the corral frightened and bewildered.

Parking the pickup, we headed with eager anticipation toward the fence. I wanted to touch one, but as soon as we approached, the animals ran from us. We then followed the sign pointing to the office. As we walked through the doorway, an older, tired-looking man with a dirty straw hat greeted us and drawled, "How can I help you, ma'am?"

"I'd like to pick out a baby burro for adoption," I said as he motioned us toward another door.

"Well, the fellows out back will help you load up whatever you can catch."

Jan's and my eyes met in wonder. *How could we ever corner one,* we both thought in unison. Those burros were wild! Fortunately for us the men were willing to lasso and load the animal if we would make a choice. I picked out several prospects and they finally captured a yearling jenny. Although it was not one of the babies that I had chosen, by then it didn't matter. The men had their ropes tight around her neck, putting on her head the halter we had given them.

After I backed the trailer up to the loading chute, the two young men now had the challenge of getting her inside it. I could not watch for I thought for sure they would break her delicate legs before they man-handled her into the horse trailer.

The trip homeward was uneventful. Jan and I quietly contemplated what I had gotten into with my desire to rescue wounded and abandoned creatures. *If it took four good-sized men to finally load the burro*, I wondered, *how would we ever get her out?* I had paid fifty dollars to the Animal Protection Society to set her free and now I felt like the one who needed protection. She was wild and I didn't know how to tame her.

When we arrived home she stood motionless in the trailer as we opened the door, staring at us with suspicion. Jan, who had more courage than horse sense at that moment, was first to approach her. She untied the lead rope and handed it to me. I gripped it, ready for anything. The burro turned her head from side to side, glanced through the open door, and to our surprise, just walked out of the trailer and down through the pasture gate as if she knew she was home. Little did I know then how she would respond to the farm animals.

I named her Jubilee in allusion to the fifty dollars that I had paid for her and the Biblical concept of the jubilee year. She attached herself to me but her past abusive treatment made her terrified of men! She also killed my chickens by stomping on them with their front hooves. She chased dogs and bit my sheep. Jubilee was wild, untameable, and worthless as far as everyone was concerned. Love would win out, I kept telling myself and anyone else who would listen to me. Most asked, "Why don't you get rid of her?" But she was beautiful and I had fallen in love with her. Jubilee had long velvet-like ears and large brown eyes easily streaked with fear. Her coat was gray in color with white under her chest and body, and with a dark mark across her shoulders and down her back in the shape of a cross. I'd sit in her pasture and watch her, occasionally feeding her extra treats. That year when she was extremely sick with pigeon fever she came to me for help.

She has mellowed in the years we have had her. Chuck and the grandchildren can pet her now. Jubilee loves kisses on her nose, and when her foal is born, we will raise it with the sheep to be a guard burro. Her trumpet call resounds all over the neighborhood. Love did transform!

"O Lord, it is true that many have been abused and mistreated on their journey through life. As a result they suffer the longlasting effects of such brutal treatment and broken hearts. Could their wild actions of self-destruction or even self-preservation really be a symptom of their pain and past? In that case such behavior is not something to fear so much as a signal we need to listen to and respond to them in love. Oh, help me listen to and love others as You have done to me. For love will win!" ♦

Cuddles

OUR family had gathered from far and near. That summer we had to work on a fence to surround the lambing pasture our ewes would need the following spring. My husband took a week-long vacation. The "men folk" worked day after day in the summer heat, planting redwood posts in freshly dug holes, stretching wire fence to the limit.

They took out time only to stop for the large noontime meal that the "women folk" had prepared in love. Then everyone fell into bed each night, exhausted. Our busy schedule involved total dedication to the fence line, because we must keep the predators and wild dogs from hurting our unborn lambs.

After three days of our hectic pace, I realized that I hadn't seen our cat, Cuddles, for morning or evening meals. Usually she was always under foot, clamoring for her tuna. Although over five years old, she was still as soft and cuddly as a kitten. In recent months, though, she had grown detached and seemed merely to endure the affection we lavished on her. But we believed that if we loved the "un-lovable" in her she would be more responsive . . . and she was.

I began my search for her, calling as I went in and out of the house. Then I happened to glance down from an upstairs window and saw her lying by the edge of the tree. Terror grabbed me, and I raced out the door to find Cuddles dead in the front yard. How many days had she been there, unnoticed? The neighbor's dog had killed her in our very own yard while we had our attention occupied with building the fence to protect our lambs from the dogs and coyotes. I wept as I thought how no new cat or lamb could take her place.

"O Lord, are we busily hurrying about building new programs, creating new ideas of shelter for the yet unborn lambs of Your fold, and our very own are dying right in the church pews in front of us. Irreplaceable people, no two alike. Even the new convert who comes to You will never replace the one who has just perished. Help us, please, to love the unlovable, to feed and nurture those who are with us, so with the newborns we can enter together the shelter of God's love forever." ♦

MacDougle

HE looked confused, dirty, and lost. Three days in a row went by before I finally stopped. I had been too busy rushing back and forth to town as I prepared for a reception for 1,200 people—my biggest catering assignment ever. The fact that I always saw him on the street corner by the firehouse kept me alert each day to see if he was still there on my return trips.

On the afternoon of the third day, I finally pulled the car to the edge of the road and looked more carefully at the little stray. He was older than he first appeared. His long hair hung over his eyes and ended somewhere around his mouth where stickers had lodged for some time. Strangely he seemed quietly relieved that someone had noticed him. The rest of the hair down his back was matted to the skin, and mud clods clung to the long fur, and weeds hung on all four legs, causing him to flap and clank as he walked.

"Poor pup," I spoke gently as I reached to pick him up and put him in the back seat of my car. Surely he must belong to someone around here. But when I knocked at the door of each home, no one knew where he came from or cared. I drove home thankful that I was not stealing him.

For several days the dog kept falling off the back steps, blinded by his overgrown hair that reached below his chin. By Friday I could stand it no longer and whisked him off the porch to the local grooming kennel to be washed and clipped. When I went to bring him home, I didn't recognize him. I was sure this handsome, clean, bright-eyed dog was not the same animal I had brought in several hours earlier. But when he heard my voice, he gave a faint wag of the tail, and I knew he was mine.

We named him MacDougle. This mixture of poodle, terrier, schnauzer, and some other unknown breed certainly looked the part. All he needed was a little green tam for the side of his head. He enjoyed riding in the pickup and went with me on several occasions while I purchased supplies for the reception.

One morning I had to leave early to pick persimmons at a friend's ranch before it started to rain, and forgot all about MacDougle until about midday. A thunderstorm raced through the area. When I returned home that afternoon, he was gone. I called his new name, but to no avail. MacDougle did not come for supper or for breakfast the next morning.

I kept to my schedule as I planned for the reception—a job I hadn't wanted to do in the first place. Since my early morning jogs helped to ease the pressure, I tried to enjoy the beauty of the late fall leaves. Shades of gold and red mingled with the misty fragrances of dawn. The leaves carpeted the ground, having been knocked down by the wind and recent rain. The crisp breeze brushed against my face and my mind wandered in different directions.

"Where is MacDougle?" I kept asking myself. *"Why did I say yes to this high pressure event for our guest church who rented our sanctuary on Sunday?"*

Their church had been under construction and the structure had caught fire and burned. Much was lost and spirits were depressed. The guest speaker, a well-known, popular Evangelical charismatic minister who had built an impressive structure of his own, had accepted the invitation because this struggling congregation was affiliated with his church in some way. Everyone wanted to catch a glimpse of him when he arrived, shake his hand, or even get his autograph. But I just wanted to do my job and asked for God's presence to pervade the room, so everyone would know that He was in charge of the reception. I was frustrated, and I missed MacDougle.

By the end of the week I finally fell on my knees at the foot of my bed, arms outstretched over the spread. "O Lord," I wept, "I don't know why I care so much for that stupid little dog. He was only a stray, and all I did was rescue him, have him washed and fixed up, and he snuggled inside my heart. I would like to have him back or at least know what happened to him."

Then it happened, a moment of silence while I knelt there. Thoughts rushed through my brain, rapid in process, but so very gentle.

"Oh, my child, that is exactly what I did for you—rescued you, washed you in My blood, and fixed you up not by mending your broken heart but by giving you a new one. That is why I care so much—because you are engraved upon the palms of My hands."

The tears flowed and I realized anew God's mercy and love for me. If that was the reason for MacDougle, it was all right. I could handle his loss if it was only to hear God's voice. I got up from my knees willing to accept whatever happened.

The next morning Sandy (my neighbor, who has helped me in all kinds of adventure) and I had to purchase and pick up flowers. I had ordered pots of yellow mums to decorate the reception hall. As I turned the corner by the fire station, there in front of my orange pickup truck was MacDougle, with his funny little sideways gait, trotting toward me.

The truck hardly stopped before I flew out the door to sweep him up in my arms. I had found him again. Since then we have discovered he is terrified of thunder storms. He now stays close to home.

Postscript: The reception went well. The food, they said, was tasty. The yellow mums sparkled in the candle glow. There was enough hot apple drink steeped with cinnamon sticks for all. The Lord's presence was there. And the guest speaker had such a tight schedule that he had to rush to catch a plane after he spoke, being unable to attend the reception given in his honor. ♦

The Mare

FRIDAY'S usual routine beckoned: cooking, cleaning, and preparing the house for Sabbath.

Earlier that morning, I had returned a telephone call about a mare I wanted to see. She was a gray Arabian and had all the qualities I'd been looking for. Possessing good blood lines, she was reputed to be an excellent producer, and even promised to be a good trail horse. While the price was high, I figured that the terms were negotiable, and somehow I would work out the financial arrangements later. It might be a good option to lease her.

I put my pen on top of the directions I'd written down and sighed. How could I make it out to Lincoln, twenty miles or so each way, and still get my house cleaned by the time the sun set? Rationalizing to myself, I figured that it would take about one-half hour to drive each way, and twenty minutes to look at the mare. The trip would total, at the most, an hour and a half. Since I'd still have time to get everything accomplished, I decided to go.

As I walked through the house, however, I entered the library where my Bible, reference books, and study guide lay scattered on my desk. The thought flashed through my mind, "What am I doing? This is Friday, and I haven't even finished studying the material I must teach my class tomorrow." Then my mind skipped to another thought. "I'll give myself half an hour to finish the Friday portion of the lesson, put the presentation all together, hurry on about my horse adventure, and still manage to get everything done by nightfall."

While I was studying the "Think It Through" question on the lower left hand of the page, a sentence leaped out and landed right in my conscious mind: "In view of the glorious heavenly future, what should I be doing now?" I knew the voice of the Lord had spoken. With pen in hand, I answered slowly on the pages of my quarterly: "Preparing myself, by God's grace, for He is preparing a place for me."

Yes, it all fit so perfectly. Today was Friday, the preparation day, and I had almost forgotten. So eager was I to go and look at an $8,000 mare that I had let the desire overwhelm me. I had been studying and writing notes all week, working on concepts of what effect our trust in Christ and His soon return had on our personal daily habits, family relationships, use of money, and the use of leisure time. My love of horses had blinded me to my own need.

The lesson was for me! Needless to say, I didn't go on my planned outing for the day. By God's grace, my heart and home were spotless and aglow as the setting sun cast the last shadows through the trees in the back pasture. Another precious Sabbath had begun.

Numerous blessings came my way the next day as I taught my class. I'll long remember the ordinance service, lunching with friends, and transferring everyone out to our place by car that afternoon. As we concluded that Sabbath day with a quiet walk in the woods, watching wild geese float across the golden waters of sunset, my vesper prayer was this:

"Father, I can hardly wait for the ultimate Sabbath You leave prepared for all who love You, and I long for that glorious day when Jesus returns. Help me use wisely these final hours of preparation in His dear name, Amen."

Postscript: Several months later after I wrote this story, a friend of a friend gave me a beautiful gray Arabian mare. ♦

OTHER PASTURES

"I have other sheep that are not of this sheep pen. I must bring them also. They too will listen to My voice, and there shall be one flock and one Shepherd"

(John 10:16).

Backyard Goat

THE barking sound awoke me as I lay sleeping
there.
I rolled on over to ignore, my mind said,
"Be aware."
Then I listened closely, a wee, small cry was
heard,
I leaped up from my bed, I didn't say a word.
Rushing to the window to hear the noise again
I realized with horror it was coming from a pen.
Yes, in my own backyard where my goats were
penned up safe,
I heard the dying sounds of the wounded
little waif!
I ran out in the night so dark with flashlight
in my hand,
Grabbed a pitchfork on my way to fight off
the roaming band.
When the light shone on them, the dogs they
fled in fear,
I said within my heart, precious Lord, I
pray, be near.
Now I rushed on to my goat and knelt and
prayed,
"Lord, help me with this animal so wounded and
afraid."
Suddenly I realized I had not called my husband
dear,
I ran back to the house and cried with voice

of fear,
"The dogs have tried to kill our goat, awake,
    awake, awake!
We must hurry back again for the broken,
    wounded sake!"
We dressed and hurried out into the darkest
    night,
I continued praying, "Lord, help us in our
    plight."
The vet was called, directions given, and
    blankets were brought down
To wrap the wounded animal, as he lay dying
    on the ground.
The morning light came slowly and the neighbor's
    truck appeared
To take us to the vet to hear the words we
    feared.
"I do believe there's hope," he said, "I think
    he'll be all right,
He will have to want to heal and for his
    life to fight."
I had bitterness deep within, for that dog had
    been the cause.
The dogs had roamed both far and free, not
    regarding canine laws!
Then my neighbor said, "I am sorry,
    please forgive."
Now Goatie's up and walking, he regained his
    will to live.
He was just a little creature in the backyard

of my life,
But this taught me oh so much about evilness
and strife.
How the mad dogs of this world are eating some
of us alive,
We must have listening ears or we all just
won't survive!
For the dogs have tried to kill our friends,
awake, awake, awake!
We must hurry out to rescue, please awaken for
their sake! ♦

The Sacrifice

THE tailgate of the pickup slammed shut. Dust settled around our feet. We had loaded the white wethers, and I stood there numb in the aftermath of my first sale of locker lambs. This had not been my reason for raising natural colored wool sheep. But the bargain had been made, and I had to hold up my end of it. Wanting a soft lamb pelt for a bedridden friend, I had exchanged two live lambs for three pelts. One I'd receive now, the other two when they were processed, after the butchering. *With every gift of love there is a sacrifice,* I thought, as I stumbled up the pathway to the house and the truck drove away.

Eager to give the gift, I traveled to her home. When I arrived with my newly-acquired soft white pelt, I found to my dismay that she was allergic to wool! My mind raced with jumbled thought. I had sacrificed a lamb for this gift of love, and she couldn't use it because of her allergies. Now I couldn't bear to tell her one of my own pet lambs had died to make the gift possible. True, the pelt I brought her was from another lamb, but because of the trade I had made, it might have been from one of my own.

I prayed inwardly that her allergy could be healed so she could enjoy the benefit of the gift, and all the while God's voice kept saying in the recesses of my mind, "Oh my child, that is what I have done for you, a sacrifice of love in My very own Son who died so you could live. Please let me heal you of your allergic reaction called 'works' and let Me share the robe of righteousness I have provided for your soul."

When I left that day my friend rested comfortably on her woolen bed. Her body was relieved of discomfort, and I anew had accepted the gift of salvation.

It is true that with every gift of love there is a sacrifice. When I accept with love God's gift to me, I sacrifice self at the foot of the cross. ♦

Miracle

OUR usual morning exercise routine was to run past a small farm bordering our property. For most of us in the area, raising horses was a hobby, and Barb's mares and foals were outstanding. Especially Miracle, a yearling filly with black flowing mane, luminous warm dark eyes, a shiny coat, and a soft velvet nose. I loved her!

As we ran past the white rail fence that day, I noticed all the horses were grazing on the lush green pasture, all except Miracle. She was down, and when I called her name, she didn't move or flick her ears. Sensing something terrible had happened, I climbed over the rail and rushed toward her, while Sandi, my jogging partner, raced home for help. By the time I reached Miracle, I knew she was dead.

My legs turned into unmanageable sticks, and I fell in the cool green grass beside her. Deep sobs welled up within my throat. Miracle was dead and gone forever! How could I ever tell Barb, who was away with family on vacation?

"O God," I started to pray, "why Miracle?" My mind wandered back in time. . . .

Prayer had been the beginning of the horse's life. The morning Miracle was born Barb had called and asked me to pray because her mare was having problems. She called the vet to check the long-overdue foal.

"I'm afraid she's lost it." He removed the stethoscope, shaking his head. "Bring your mare down to the clinic and we'll try to save her."

They loaded the large bay mare into the trailer and rushed down the narrow dusty road to save her life. I prayed. While removing the foal, the vet discovered that it was weak but still alive. A small delicate filly lay limp but breathing, on the soft rice-hull bed. Mare and foal would survive.

Several days later when Barb's pickup was being repaired, she called to borrow my truck, and we hooked her horse trailer to it. Eager to see the filly, I drove and we chatted about the filly's new-found strength. That was the day Miracle came into my life.

"What are you going to name her? Miracle sounds like a good name to me," I suggested in eager anticipation of Barb's acceptance. My friend recalled the pedigree name and said they would probably use that.

A few days later found me at the fence line asking about the health of the new filly. With a twinkle in her eyes, Barb waved her hand as she pointed over in the direction of the mares. "Oh, Miracle is doing fine." We both smiled as we hugged each other, unaware of the wooden fence between us. She often commented, "If I ever sell her, I promise to sell her to you."

From my side of the pasture fence I watched Miracle grow. She was a beauty with long straight legs and loved to run.

My thoughts came abruptly back to the present —*run!* That is how it must have happened. I pulled my flying thoughts together and stumbled home.

I knew Tom, from a neighboring farm, was caring for Barb's stock while she was away. After several calls I put the pieces of the tragic puzzle together. Tom had discovered Miracle's broken leg when he went to feed the animals on Thursday night. He summoned the vet who quickly determined that she couldn't be saved. They had called Barb and obtained permission to put the filly down. I felt much relieved. Miracle had not died alone nor had she suffered all night as I had originally thought. The vet put her down in the most humane way possible. But she was gone forever!

The catastrophic thought of eternal separation flashed through my mind that fall morning, and haunted me for days. I remembered a precious friend running wild, and I wept in prayer.

"O God, You went out to feed Your flock early morning after morning and she could not come. You searched for her in the cold, dark pasture of life and found her broken, unable to stand, and You loved her so. Help me to share with others that You alone have the power to mend broken hearts, minds, and bodies. You can restore health to the most deadly illness and have the power to resurrect the dead both physically and spiritually.

"Please give her strength anew to choose You for healing and restoration, so You won't have to be her Merciful Veterinarian, who will put her out of her unbearable pain forever when You come. Let her again seek You as her Great Physician. Lord. Don't let it be like Miracle with her—gone forever. I can't bear the pain in my aching heart. It swells into deep, choking sobs. O God, please make another miracle, a lasting one. One that is not just a name given a beautiful filly You endowed with life for a short time, but the kind of miracle that happens in her heart and lasts throughout eternity." ♦

Pack Trip

DUST swirled around our horses' hooves as their horseshoes clicked on rocks. White lather seeped from under the saddle blankets. Several horses had child-sized bed rolls tied to the backs of their saddles, and food and clothing stuffed in saddle bags. Our long-awaited, overnight pack trip with the horses had begun.

A small oasis on the trail where water would be abundant and where we'd find a safe place to sleep and tie the horses was our destination that day. We stopped enroute to swim in the lake, unsaddling to allow the horses a chance to swim.

Afterwards, we took to the trail again, crossing near the area where two rattlesnakes were devouring a rabbit. One snake veered off under the horse ahead of me. The rider most fearful of snakes passed safely unaware of the danger. Somehow, as we noticed more snakes, we discovered we'd left the trail. Retracing our steps, we made it back safely to the trail we were supposed to follow. When we made up our camp that night, we were thankful for the safe journey with three children, two women, and five horses.

Morning came all too soon. My mare had developed some saddle blisters from a borrowed saddle, and I couldn't ride her. Adjusting the cinch to what I felt would be enough pressure to keep bedroll and bags safely attached, we began our journey homeward. I had to take the lead on foot to set the pace for the rest of the horses, or else

Shenda and I would have been left in the dust. Shortly after leaving our campsite, we had to travel over a narrow trail with high rock on the right, and a deep drop off to the lake below on our left. I wanted to hurry past that unsafe area as quickly as possible.

While trying to keep my attention focused on the trail ahead of me, I did glance back quickly and was stunned to see that the saddle had shifted on Shenda. To my horror, I saw it hanging from her left side, the saddle bag dangling under her belly, and the bedroll hanging over the precipice. Jan, who was at the back of the pack and the better horsewoman of the two of us, quickly took charge.

While I struggled with my cinch to readjust it, she kept the kids and horses from panicking on the ledge. Speaking with firm control, she had everyone dismount and step around with reins in hand to the head of their horse. "If anything is to go over, it will be the horses and saddles, not the children," she called. Everyone stood still. Jan slipped past all the horses and riders to come to my aid. There was no way I could pull or push my saddle back to its original position. Shenda stood motionless while I prayed inwardly for wisdom, strength, and calm. Finally, between the two of us, we restored the saddle to its proper place, tightened up the cinch again, and quickly walked everyone past the precipice.

"O Lord, on my journey homeward to Your eternal heaven, may the path, though narrow, be safely traveled. May I keep my eyes straight ahead, not looking to the cliffs on one side or the deep gorge to the waters below, where I might fall to my death. May the cargo I carry—the message of God's love to a dying world—not slip from me when I am tired and sore. May the church where I worship not cling to programs, concepts, and theology, good as they all are and much needed. If anything must fall over the precipice, Lord, let it not be Your children. May we see Your footsteps on our journey's path, and please lead us safely home." ♦

The Fold

THE small chapel was sparsely populated with gray-headed men and women, some appearing older than their years. The stress and trauma of life had made them that way. Their eyes, however, were aglow with the love of God. As I stepped into the back row for the meeting, I was astonished to see no one I knew. The lost sheep was returning to the fold, and she recognized no one there.

Where were my friends? All those I saw in church each week, those with whom I had gone skiing on holidays, had dined with at the fashionable restaurants after church, had associated with on social occasions? Did no one I know attend the mid-week prayer services? The meeting was pleasant, and as we bowed in a circle of clasped hands, my new-found tender heart broke with loss.

Oh, how I missed you tonight, my friend,
You I have loved for so long.
Did you think of me saying a prayer for you
As we knelt in a circle to pray?
Were your thoughts so consumed by the cares
    of this life
That you hadn't seen time slip away?
How can I say it?
The words make me weep.
Jesus still loves you
So many asleep!
"Wake up!" shout I,
Will they all pass by?
Dear Lord, in Your mercy
Bring all Your sheep home,
Those who have wandered so far from
    the fold
Don't let them stay out in the darkness
    and cold.
My shepherd You've been and I look to
    You now
To show me the way, to show me how.
How can I share all the love in my heart
The love You have given me right from
    the start.
May somehow I be just a channel of light

To others I meet in the dark of the night.
But, oh how I missed You tonight, my friend,
You I have loved for so long. ♦

PEOPLE PARABLES

"Know that the Lord is God. It is he that made us, and we are his: We are his people, and the sheep of his pasture" (Psalm 100:3).

The Closet

MANY months before, we had purchased the round-trip tickets so there would be no possibility of our children not getting a flight out. The holidays had arrived, and our Alaskan family had made it home safely.

Our house had started to look festive. From outside, the tree was beautiful, the small delicate gold and white lights sparkling and glistening through the window pane as if moved by a soft breeze. The inside of the house was warm and toasty, made that way by love, laughter, sharing, and glowing embers in the fireplace. Christmas was a-coming!

The children had gathered from snowy climes to spend a part of the holidays with us. Visiting friends, neighbors, and relatives came from distant places. The beauty of poinsettias complimented the smells of homemade candies, pies, cookies, and boxes of mandarins. Christmas music played on the stereo as handmade gifts— some given from real sacrifice, but all from hearts filled with love—were secretly wrapped. Gifts, the symbol of Jesus, abounded. One surprise gift came un-wrapped as a grandmother witnessed her grandson take his first unaided steps from one parent to the other.

As I gaze back over that holiday spent with family, one lesson stands beautifully reinforced in my mind. My daughter was looking in my closet for a pair of shoes as she said, "Mom, I knew I didn't need to bring any shoes, for I knew I could wear yours." I'm so thankful she knew me well enough to realize that I would share my closet with her, even down to the shoes.

"O, Lord, you have a closet full of robes, a size to fit each one of us. And the best part of our journey home will be the one-way ticket. No more goodbyes! We will say with hearts full of love and child-like faith, 'Father, I knew I didn't have to bring a robe for I knew I could wear Yours.'

"Praises forever for the best present ever given: JESUS our REDEEMER and KING!" ♦

Phone Call

WE had just said the blessing on our breakfast meal that Wednesday morning when the telephone beckoned. Saying, "Good Morning," in my usual fashion, I was suddenly brought to the reality that it was not the good morning I had expected.

Steve, my son-in-law, spoke long-distance words that made my knees buckle, sending me into the nearest chair. Dawson, my grandson, had acute lymphocytic leukemia.

Thoughts raced frantically through my mind. My sight blurred with tears. I realized Steve needed strength, not weakness, at that moment, and my voice steadied.

We made flight arrangements and met the sad little family at the airport that evening. Then we rushed the three tired loved ones home for a short night's sleep, for the next day they would admit Dawson to Stanford Children's Hospital.

Our eyes opened to a whole new world our family had never seen before while we lived in the hospital those first weeks. Death and sickness were on every side. Children with hairless heads looked like something from a science fiction movie. I prayed we could endure.

God answered our prayers as He helped us keep our eyes focused on Him and not on those around us. That freed us to reach out in love and sympathy to the afflicted and their extended families. Meanwhile, the prayers of those who loved us from across the land mingled with our own petitions traveling heavenward to our heavenly Father, who says, "Ask and it shall be given. . ." Doors opened so his parents could transfer the child to Davis Medical Center, and we came home. His hospital stay was short and he was home on the next to the last Sabbath of the month.

Home at last suddenly felt like a great privilege as everyone was again under one roof, with no more sleeping schedules at the hospital. We could now restructure our lives. Dawson was anointed that Sabbath afternoon, and we were all filled with God's perfect peace. The week went quickly by, and my daughter Karen used the time to rest before she gave birth to Joanna the following Sabbath afternoon. Thank God for a sacrificing doctor and a quick delivery. Mother and baby came home the next day. Our lives were filled with thanksgiving and praise.

Today we look at each day as it comes, something we learned from our first encounter with tragedy. To survive we must do this. But through it all we have acquired trust in God. Reflecting over the past, one of the lessons so beautifully illustrated for me was when Dawson would take his bitter medicine without complaining because, cradling him in love, his mother taught him to obey as a wee infant.

I thought, "O Lord, how much You would like to see Your children in afflictions swallow the bitter pills of life without murmuring, to trust in You, Heavenly Father, knowing You wouldn't harm us. And when the pill is swallowed, You encompass us in Your loving arms and whisper, 'I love you, my child, please trust me.' Help me be like that, O Lord, I pray!" ◆

Blanket of Snow

ON an early March morning in 1985, the snow fell in a soft blanket on the Sierra foothills, covering the earth with a garment of white. Our grandson would be buried that day.

"What weather," I sighed as I reached for the ringing telephone. A friend had called to tell us that the authorities had advised all drivers who ventured out to use snow chains.

We left early in the heavy rain so we could drive to Penn Valley to pick up our daughter and son-in-law enroute to the church in that community. We traveled with great care as the rain turned to snow and we made our way to higher altitude. The storm continued, making travel nearly impossible. As we rode along we saw cars filled with our friends and loved ones, traveling in the same direction, moving slowly, steadily. The other drivers, determined as we, headed uphill through the falling snow. Some people turned around for fear the storm would get worse because no one had expected the snow to fall so heavily and none carried chains.

"I'll crawl over these mountains to be there" I said to myself. Then I prayed, "Lord, please help each one to get there safely."

My sister arrived after the service ended, while we were in the church foyer accepting the love and warmth of all our friends who had made the long pilgrimage in the ugly weather to share in our day of sorrow. Her disappointment was evident.

"How did you make it through the storm?" I whispered, knowing Interstate 80 was closed.

"I took the Feather River route. I would have started earlier if I'd known it was going to take so long." We exchanged hugs and shed tears. Shortly after the funeral the storm abated, and by the time we arrived at the cemetery, where we held a short grave-side service for family members, the sun had pushed through the clouds with golden streams of light. There was hope in each ray of sunshine. God was in control and with us as we laid Dawson in his resting place under the blanket of snow and flowers to await that glorious day of Jesus' coming.

Time has illumined that day for me, and now I can pray: "Lord, please let all of us who attended Dawson's funeral that sad day, and those family and friends we represent, be ready for that happy reunion in the sky. Don't let anyone turn back when the snow falls hard. Let us always carry the 'chains' of Your Holy Word which can cut through the slippery ice on our narrow path. May there be a 'Feather River Route' for each one of us when the storms of life block our way to our heavenly destination.

"Will it take longer than expected for some of us to come to You? Should we start a little earlier because the detour is long and difficult? We all must be there for that grand and glorious day when the Son breaks through the angel clouds and the sky will be filled with golden streams of light . . . 'and the dead in Christ shall rise first. Then we which are alive and remain will be caught up together with them in the clouds to meet the Lord in the air; and so we shall ever be with the Lord' (Thes. 4:16, 17)." ♦

The Celebrity

IT all happened after my picture appeared on the front page of one of the local metropolitan newspapers. The article included a photograph of me holding a gorgeous gray fleece, standing in my west pasture amid spring poppies and lupine. Different pictures and other stories appeared in the smaller suburban and foothill papers.

I had become a celebrity of sorts and discovered this status made me sought after to offer advice and share information on the raising of black sheep. Invitations to demonstrate spinning skills poured in from 4-H groups. People expressed their desire to learn my opinion on types and textures of wool, methods of breeding, and farm management techniques. Individuals wanted to buy sheep from me and sell me sheep. Others wanted to discard their old sheep in my pasture or their old wool in my wool boutique. Phone calls came by the score, and I began to think I needed a secretary.

Once again the phone rang. I lightly joked with a stranger on the other end of the line, a voice I didn't recognize. He introduced himself as the host of a popular magazine-style television program and, at first, I thought someone was playing a joke on me.

"How did you learn about me?" I asked, when I finally believed he was who he claimed.

"I saw your picture on the front page of the *Union* and thought your story might make a good spot in our program . . . you and the sheep and all."

I guess that's what happens when your sheep-raising friends ask you to be the president of the local sheep-growers group you helped start. Sierra Shepherds was fun—we were meeting our goals. We had become known as a responsible "loosely-knit" (no pun intended), but well-organized group of shepherds, breeding and raising colored sheep. As a result we sold more wool and sheep collectively than we ever did individually.

The taping lasted over two hours and went well. I asked my daughter and granddaughter to make their television debut with me. Several months later we watched *Evening Magazine* on a Thursday night and saw three-and-a-half minutes of beautifully edited film featuring the wool of Sierra Shepherds and my sheep on our farm. It included a short interview with Karen on spinning. Then the television host interviewed me regarding farm life and I gave a short demonstration on the shearing process. The interview closed with Joanna feeding live sheep.

I felt uncomfortable in my brief encounter with fame. I'd done volunteer church work for years and never received such applause. Parts of me reveled in the clapping, but deep in my heart I knew God had something else for me to do. "Feed my sheep" and "write a book" He had said some years before, and here I was going around the county proclaiming the good news about raising colored wool sheep. I had been having good, clean, wholesome fun, but I had another destiny. My hour to decide had come. I felt somewhat like Moses must have. The choice was mine to make—the pleasures for a season . . . or to suffer affliction with the people of God. Moses chose the people.

"O Lord, I'm so glad You helped me choose the people too. You gave me the strength to make that decision. While You know how much I love my lambs in the pasture, I love Your human sheep even more. You were there by Your spirit that night I resigned from the presidency of Sierra Shepherds. You gave me the strength to stand firm to my convictions. Help me to remember not to applaud, but to faithfully affirm others as we work together for the good of Your flock.

"Please keep in mind that I am a novice shepherd and need Your presence when I am called to speak in public places, give lectures, sermons, or teach my class at church. May I not forget the Sheep Manual (Your Holy Word), and read it faithfully, for only from that perspective can I truly share the Good News—salvation is free and God is love." ♦

Greener Pastures

SHARED experiences of pain and prayer are some of the ways God brings His sheep together. In this book I have shared portions of my heart with you. My life will have other prayers, pastures, and parables. Some I will write about and others will remain my own. But for this moment I want to share my deepest desire for you.

If you have not already discovered what a pleasure it is to be one of the Flock, or if you have wandered from the fold, or if you're part of the Flock but have lost some of the joy of living, may your heart be lifted up to gaze upon the Lamb of God which takes away the sin of the world. May you hear the Shepherd's voice in such new dimensions that you will be astonished beyond belief at what He has in store for you. And may you hear Him call your name and find that place of rest in His green pastures.

A number of years ago I took a class in an art form called decoupage—a creative way to use scissors, paper, and glue. When I completed the class I was awed at how fertile my mind had become, once stimulated. For days thereafter all kinds of pictures in magazines, seed catalogues, calendars, and greeting cards caught my attention. Illustrations of any type became a virtual playground for exercising my new craft. I hardly had enough time to cut out all the artistic pictures I collected.

Now my prayer for you is that by the beauty of His Spirit you will be stimulated to hear His voice wherever the Lord chooses to feed you: at your desk, amid the flowers of the field or a green glade, as you view a sunset or the stars of the heavens, at the end of a bed where you kneel to pray, in a bubble bath, at the kitchen sink or in a garden, gazing out the window of a car, on a flight through billowing clouds, at the office or in an operating room—anywhere.

May you gain a verve for life that sparks spiritual comparison in all the myriad of daily activities that are unique to you. *For there you will find your personal pasture.*

Listen to Him. Bask in His forgiveness and mercy. Delight in His intimate love and care. Our great Shepherd is coming to take us home to that heavenly grazing land, where the lion and lamb shall feed together. He has already given His life for His sheep. It is up to us to accept the greatest sacrifice ever made.

Please make a date with me to meet in those greener pastures. "My sheep hear my voice, and I know them and they follow me" (John 10:27). ♦